The Heart Brain

Did you know you have 3 brains?

OTHER BOOKS BY CATHERINE ATHANS

Make Your Dreams Come True Now

CD: *Make Your Dreams Come True Now, Guided Meditations*

Just Imagine

Truth or Consequences

Rainbows are Real

U Just Do It!

Breathe

The Heart Brain

Did you know you have 3 *brains?*

~~~~~~~

Marie-France Louvel

Catherine Athans, Ph.D.

ANGELS ISLAND PRESS
Los Altos, CA

Copyright © 2011 Catherine Athans, Ph.D.

Second Edition 2018
978-0-9983575-3-9

Angels Island Press
An Angels Island Production
303 First Street
Los Altos, CA 94022
www.AngelsIsland.com
1-650-948-1796

All rights are reserved. No part of this book may be used or reproduced in any manner whatsoever without prior written consent of the publisher except in case of brief quotations embodied in critical articles and reviews. Special book excerpts or customized printings can be created to fit specific needs.

Library of Congress Control Number: 2011906251

Publisher's Cataloging-in-Publication
(Provided by Quality Books, Inc.)

Louvel, Marie-France.
　　The heart brain : did you know you have three brains? / Marie-France Louvel, Catherine Athans.
　　p. cm.
　　Includes bibliographical references.
　　ISBN-13: 978-0-9794380-2-8
　　ISBN-10: 0-9794380-2-0

　　1. Heart--Diseases. 2. Brain. 3. Stress (Psychology) 4. Mind and body. I. Athans, Catherine. II. Title.

RC682.L68 2011　　　　616.12
　　　　　　　　　　　QBI11-600090

DESIGN: DOTTI ALBERTINE | WWW.ALBERTINEBOOKDESIGN.COM
IMAGES: THINKSTOCK/GETTY IMAGES

*To All Hearts
All Over the World*

CONTENTS

Foreword		*ix*
Preface to Second Edition		*xi*
Preface to First Edition by Dr. Catherine Athans		*xv*
Introduction by Marie-France Louvel		*xix*
CHAPTER 1	THE HEART'S DESIRES	1
CHAPTER 2	THE TWO BRAINS IN THE HEAD AND THE HEART BRAIN	5
CHAPTER 3	HOW THE TWO BRAINS IN THE HEAD WORK TOGETHER	13
CHAPTER 4	THE IMPORTANCE OF THE HEART	21
CHAPTER 5	THE BRAIN IN THE HEART	31
CHAPTER 6	EMOTIONAL MASTERY	43
CHAPTER 7	EVERYTHING STARTS WITH EMOTIONS	49
CHAPTER 8	THE ELECTROMAGNETIC FIELDS OF THE HEART	53
CHAPTER 9	THE LANGUAGE OF THE HEART	67
CHAPTER 10	HEART HARMONY	75
CHAPTER 11	HEART HARMONY QUIZ	79
CHAPTER 12	HEART HARMONY EXERCISES	85
CHAPTER 13	HEART HARMONY CONCLUSIONS	99
CHAPTER 14	FINAL CONCLUSIONS	103
Acknowledgments		*107*
Endnotes		*109*
Glossary		*113*
References		*117*

FOREWORD

During the embryonic stage, before your brain was created, you were all heart. The brain formed from the heart, not the reverse. In Chinese medicine the heart is a bridge between the mind and the body. In most ancient societies creation begins with the heart, not the mind.[1]

Neuro-cardiologist, Dr. J. Andrew Armour[2] introduced the "heart brain" concept in 1991. His work revealed that the heart has the ability to act independently of the cranial brain, i.e. to learn, remember, feel, even *sense*.

When you manifest the world of form from the heart instead of the mind, you are accessing limitless power. Literally, whatever your "heart desires" is within your reach. Brain-based creativity is possible but comes with a price—you will also get what you don't want. Heart based creativity, however, is limitless, loving and kind.

In this book you will learn to regain a close relationship with your heart, listen to your heart and receive the bounty of your efforts in ways that at the moment may seem unimaginable to you.

PREFACE TO SECOND EDITION

With the advent of new scientific discoveries, we decided to revise The Heart Brain to include a chapter on the magnetic fields of the heart. The new data is so startling and important that we want to share what we have learned with all our readers.

The inspiration for this new chapter came as science developed new technology to better measure magnetic fields in the body and especially from the heart. It has become clear that the magnetic fields of the heart can be felt not only from five feet away or ten feet away but literally can be felt in space in different galaxies. For instance, Dr. James L. Oschman, author of two books on energy medicine, concluded that the magnetic fields of the heart actually extend indefinitely into space.

These revelations, along with the work of the late Dr. Masaru Emoto, doctor of alternative medicine in Japan, illustrate that by sending love from the heart across a geographical space one can literally change the shape of water crystals. The significance of this research is that it opens the fields of study of the magnetic forces that are within each one of us. These natural forces give us the ability to make positive changes in our world.

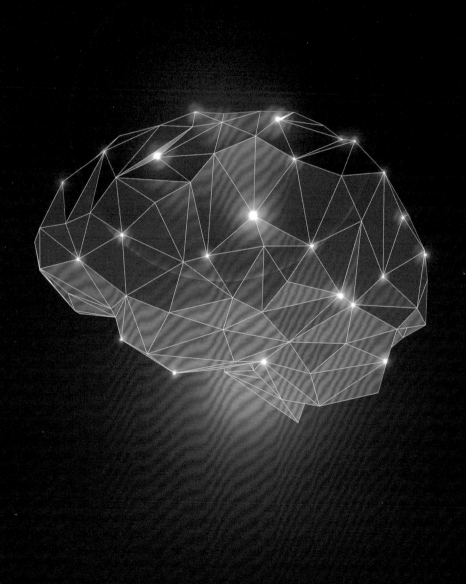

The HeartMath Institute has continued their research into the effects of love sent from the heart. Some of their findings illustrate that love verifiably changes the atomic structures within other humans, plants, and animals. In the past, this has been considered science fiction. We have now learned that this is actually scientific fact.

In contemplating this new information, we must change the way we think of the world. In the nineteenth and twentieth centuries the world was seen as a solid three-dimensional space of length, width, and depth. With the discovery and study of the effects of magnetic fields, we now know that the world is multi-dimensional. Moreover, it has changed from a world of solid into a world of flowing energies. At the heart of these energies are magnetic fields, which affect us physically, mentally, and emotionally. We are at the forefront of discovering the power that resides inside the human heart. In the near future we will use the science of love and magnetic fields to heal disease, cleanse toxic environments and promote the healthy development of all living things.

Marie-France Louvel
Dr. Catherine Athans

PREFACE TO FIRST EDITION
By Dr. Catherine Athans

"LISTEN TO YOUR heart" is a phrase that is told to everyone at some time in their lives. But what does this really mean?

Most students are taught about the importance of the brain. Science has spent a great deal of time, money and effort to map the different regions of the brain, to learn how it functions and to discover areas where consciousness is developed and supported.

However, it is only recently that science has turned its attention away from the traditional brain in the head to a new brain found in the heart. **IT HAS BECOME CLEAR THAT THE HEART PLAYS A VITAL ROLE IN THE WELLBEING, FUNCTIONING, AND BALANCE OF THE BRAIN.** So when someone says, "It is important to listen to your heart," that phrase takes on a far greater meaning.

The Institute of HeartMath took the "revelation" a step further: This institute proved, through scientific research, that negative emotions create jagged heart rhythms (incoherent-heart-rhythm) and increase disorder in the nervous system. Why is this important? Because when the heart is beating irregularly, it actually shuts down the production of many essential hormones and allows the various organ systems in the body to become imbalanced. It also disconnects its communication from the brain in the head. As a result, the brain does not receive vital information about the body, causing a slowdown in the body's immune systems.

In contrast, **POSITIVE EMOTIONS CREATE SMOOTH AND REGULAR HEART RHYTHMS (COHERENT-HEART-RHYTHM) AND HARMONY IN THE NERVOUS SYSTEM.** These regular waves are so important to the functioning of our being and our body that we have included in this book an exercise that facilitates heart-rhythm harmony. It is so startling in its effect on the whole body-brain system, yet so simple in its delivery, that anyone can learn how to practice this exercise many times a day no matter where they are. **IS IT TOO GOOD TO BE TRUE THAT THIS PRACTICE BRINGS HARMONY AND OPENNESS OF COMMUNICATION OF THE HEART WITH THE TWO BRAINS IN THE HEAD?** In reality, through scientific research (controlled testing) it is proven that these exercises bring all the organs of the body into a very communicative state with the heart in just a few short minutes.

Other bodily systems such as the respiratory system, the digestive system, and the immune system, synchronize to the heart-rhythm; harmony in the heart-rhythm leads to better problem-solving mentally and emotionally.

So, when one thinks of the heart, it is neither a mere mechanical pump, nor the strongest muscle in the body. Rather it is the most powerful gatherer and generator of information about all the systems of the body. The heart is the most important communicator of that information to the brain.

While we certainly do not claim to be scientists, nor do we claim to have done the research necessary to prove what we are saying, we have spent many hours and a great deal of energy studying this scientific research.

We are so excited about this information and the breakthroughs it can bring to everyone who is suffering in any way—emotionally, physically or spiritually—that we decided to present it to you in a simple form. *The Heart Brain* is written in a way that is easy to read and understand. We hope you will use this book to change your life in most wonderful ways. We hope you benefit from both the information and the simple exercise(s). If you do so, **YOU WILL FIND AN EASY METHOD TO CREATE HARMONY BETWEEN YOUR HEART AND YOUR BRAIN.**

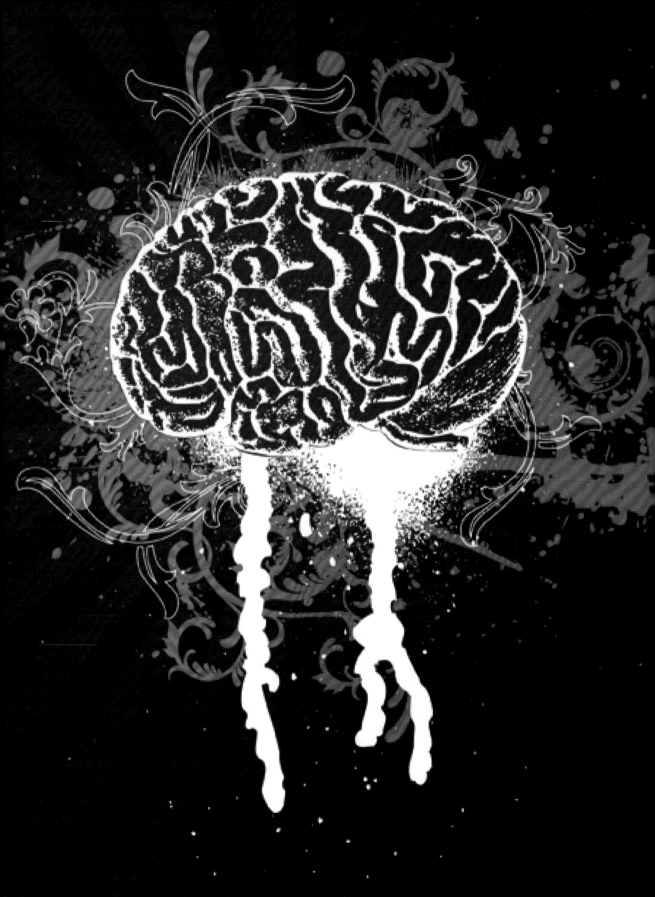

INTRODUCTION
By Marie-France Louvel

I WAS HEADING for the cash register of a Parisian bookstore, my basket full of books, when my girlfriend and counselor in French literature added yet another book. She told me that it was a "must read." I, being the stubborn and strong-willed Parisian that I am, looked at the title *Guérir (The Instinct to Heal)* by David Servan-Schreiber, MD, PhD[1]. It did not appeal to me at all. She looked me in the eye and said, "Nevertheless, you will read it one day."

Time passed and I began to have some health problems. When I went to the doctor, all he could say was, "You better do something about that stress. Your blood sugar is dangerously high, your cholesterol is out of control and so are you. What are you going to do?"

As a matter of fact, the doctor wanted me to change my lifestyle completely; to limit what I ate, make me exercise seven days a week and reduce my workload by 50%. *I didn't know how I could do this.*

I am a speech therapist and a French teacher. Many of my students have moderate to severe learning challenges. I have worked with these students for many years and care deeply for them. They depend on me. For some of these students, I am the only way that they can continue to study in a mainstream school. If they didn't work with me, some would have to enter special schools. Given the state of the economy, many parents could not support either a special school or more intensive treatment. I was not going to let those parents down!

I spent a great deal of time meditating and contemplating this situation. What could I do? While thinking about my personal situation, I became aware that probably most people feel this way about their work, their lives. There isn't time to take care of business, children and also take care of the body, the emotions and the spirit. I thought that most people have a job; they have to pay the mortgage; they take care of their kids or other people in their family and they need to exercise. Like me, all these people are tired most of the time. Even though I love my job, have a caring family and faithful friends, I am sometimes exhausted.

Not long after my visit to the doctor, I was walking by my bookcase, and the book *Guérir* simply fell off the shelf. I looked down, picked it up, and decided to read it.

I read and reread the book and found I had a deeper understanding of what stress was creating in my life. I didn't know if I could give up any of the responsibilities that were causing me anxiety, but with this book, I began to get a sense that I could change my thoughts and my feelings and gain control over the way my body was functioning. I was so excited. Could the information in this book be too good to be true?

I mentioned *Guérir* to my friend, Catherine Athans, and explained to her why I was so enthusiastic about reading it.

Catherine said, "Marie-France, translate it into English and I will read it. It sounds very exciting, and I want to know more. It could help me with many of my patients. It seems that you've come upon some very important information that should be made available to everyone."

Looking at her, I realized that she was not joking but was challenging me. I said, "Very well, why not?"

After several months of researching the information in *Guérir*, then translating and re-translating it into English, I was finally ready to give it to Catherine. To my surprise, after working so hard, I found that the book had already been translated into English under the title *The Instinct to Heal*.

As I was thinking that I had wasted my time translating the book, Catherine read my mind and explained that this was giving me the opportunity to study it thoroughly and to better understand the author's ideas.

Catherine and I decided that this revolutionary information needed to be written in a simple, non-academic format that every one could understand.

Thus, *The Heart Brain* was born.

CHAPTER 1

The Heart's Desires

Don't we all want to be happy? What does it mean to be happy? It's probably a little or a lot different for everybody.

THERE ARE PEOPLE WHO SEEM TO BE HAPPY MOST OF THE TIME. We find them in every country. They are from every social class. They are from every walk of life. Some are gifted, others are ordinary. Some practice a religion, others are agnostic. Some are married, some are not; some have children, some don't; some like money and some don't. Even though they have experienced dark moments, no one escapes from obstacles, distress, privation or mourning.  Many have overcome misfortune and achieved real meaning in their lives. They are happy. They have chosen what their hearts desire.

As you will soon learn, happiness does not come from the brain in the head. As a matter of fact, happiness comes from bodily sensations, which tell you that all is well. These sensations are gathered in the heart and sent to the brain so that you can become aware of your happiness.

Fortunate people have an open communication with their hearts. They are able to both listen and give communication back to the heart, so that the heart can send those communications to the brain.

FOR UNHAPPY PEOPLE, THE COMMUNICATION WITH THE HEART IS CUT OFF, AND THE FOCUS IS MOSTLY ON THE HEAD. These people deny the intelligence of the heart. As a result, the body and mind are thrown out of balance and cease to function at top capacity.

These people become stressed, experience sleep problems, heartburn, high blood pressure, high cholesterol, depression, and anxiety; and unfortunately their immune systems become compromised.

There is a way to establish or to re-establish the communication between the heart and the brain. It is about two special exercises with the heart, which we will teach you in this book. It will help you realize what is truly going on inside of you.

It is your heart that gives the brain information about you and not the brain telling the heart what is real.

DO YOU KNOW?

In *Webster's Encyclopedic Unabridged Dictionary*,[1] one definition of the word "heart" reads, "the human heart considered as the center or source of emotions, personality attributes … specifically in most thought and feeling; consciousness … or conscious (to know one's heart) … or : "His head told him not to fall in love but his heart had the final say."

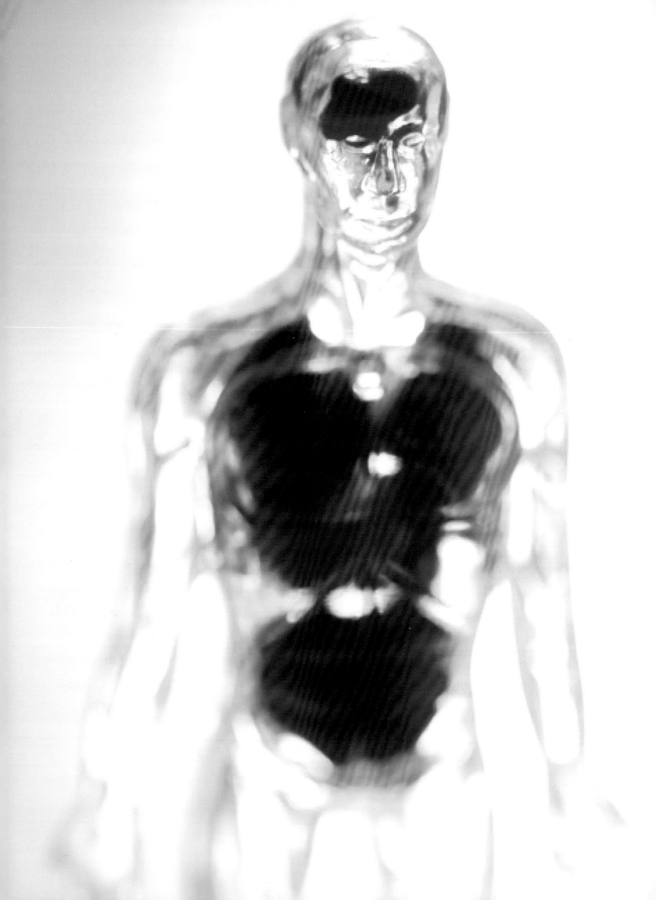

CHAPTER 2

The Two Brains in the Head and the Heart Brain

Most people think of the brain as one big mass filling up the head and controlling thoughts, emotions and functions in the body.

IN REALITY, THERE ARE TWO INDIVIDUAL BRAINS IN THE HEAD. They are different from each other. Each one has a different function. Each one can cooperate or compete with the other.

Scientists speak of a brain within a brain. They go on to explain that in the middle of the human brain is another brain. It is the vestige of our past as mammals—gorillas, chimpanzees, and gibbons. They call this the "Old Brain."

The Old Brain (also named the Emotional Brain) is made of the deepest layers of neurons. In some areas these neurons appear to be thrown together haphazardly. This brain is made up of a rudimentary structure.[1]

THE "NEW BRAIN," THE MOST RECENT ONE IN THE EVOLUTIONARY PROCESS, ENVELOPES THE OLD BRAIN IN A VERY ORGANIZED MANNER with six layers of neurons and areas that are specifically engineered for modern human life. For instance, there is an area for sight, for smell, and for sound. It is organized as a super information processing organ.

Although these two brains function almost independently, they can influence each other.

THERE IS YET ANOTHER BRAIN WHICH IS LOCATED IN THE HEART. WE NAME IT THE "HEART BRAIN."

It was recently discovered that this Heart Brain has its own network of forty thousand neurons. It has its own perceptions. Additionally, it has its own processing capacities—a hormone factory—where it is actually able to produce adrenaline and other vital hormones. It works like a conductor: it synchronizes all biological rhythms in the human being.

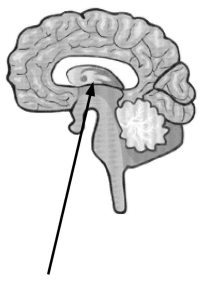

Old Brain

Old Brain

The Old Brain is a rudimentary structure similar to the ape's brain, which possesses primitive senses and emotions. It is:

- Unconscious
- Emotional
- Concerned with survival
- Tied to the body
- Self-healing
- A natural repository of emotions

The Old Brain Controls

- Body's physiology: heart, blood pressure, hormones, digestive system, immune system
- Behavior
- Everything governing psychological wellbeing

DO YOU KNOW?

When the virus of rabies attacks the brain, it affects the Old Brain, not the New Brain? This is why the first sign of rabies is abnormal emotional behavior.[2]

New Brain

The New Brain is a highly developed structure in humans that is organized for information processing. It is:

- ♥ Conscious
- ♥ Rational
- ♥ Geared toward the outside world

The New Brain controls

- ♥ Moral behavior
- ♥ Language
- ♥ Reasoning
- ♥ Elaboration of future plans

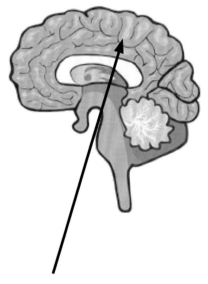

New Brain

Heart Brain

The brain in the heart has forty thousand neurons. It is:

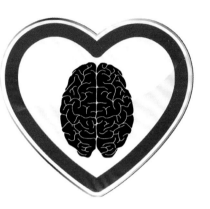

- ♥ Equipped with a small hormone factory
- ♥ Responsible for the synchronization of all bodily organ systems.
- ♥ Interconnected with the Old Brain in the head

The Heart Controls

- ♥ Information going to the Old Brain in the head
- ♥ The harmony of beats of the heart
- ♥ Blood Pressure

The Old Brain is unconscious, preoccupied with survival; and above all it is tied to the body and in particular to the heart.

The New Brain is conscious, rational and geared toward the external world.

This distinction is why Dr. Antonio Damasio, the great American neurologist and neuroscientist, tells us in *The Feeling of What Happens*[3] that **MENTAL LIFE IS THE RESULT OF A CONSTANT ATTEMPT TO BALANCE THE ACTIVITY BETWEEN THE OLD BRAIN AND THE NEW BRAIN.**

To navigate your life, emotions need to be tempered by rational thoughts. To achieve this balance between emotion and reason, *New York Times* science writer, Daniel Goleman tells us that to be happy, we need Emotional Intelligence.[4]

Emotional Intelligence is the balance between emotion and reason. Daniel Goleman does not talk about IQ, intelligence quotient, which evaluates intellectual performance. Instead, he talks about EQ, emotional intelligence, which is your capacity to identify your emotional state and the emotional state of others and to regulate your emotions in relationship to others. Life circumstances call upon you to interact with others. **YOUR CAPACITY FOR RELATING TO OTHERS DETERMINES IN LARGE PART YOUR SUCCESS IN LIFE.**

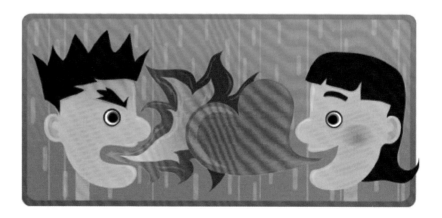

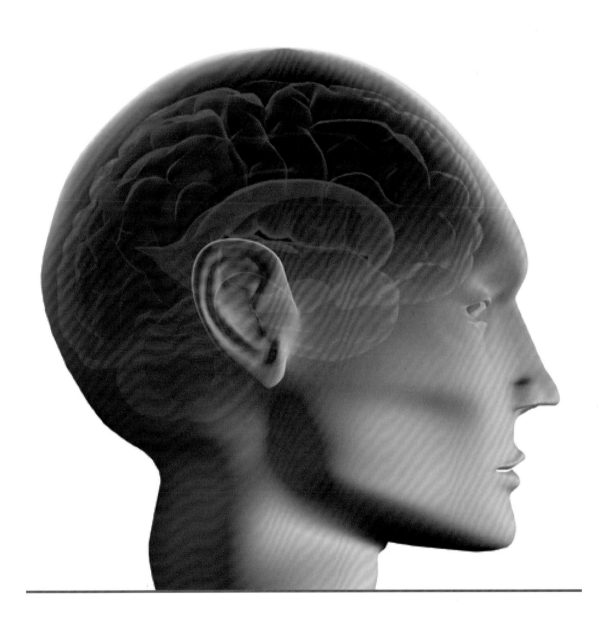

CHAPTER 3

How the Two Brains in the Head Work Together

THE OLD BRAIN directs you toward the experiences you seek, while the New Brain gets you there as intelligently as possible.

These two brains can either cooperate with each other, or they can compete. This collaboration is important because when the two brains (Old Brain and New Brain) cooperate, you are able to pursue your goals effortlessly and know what choices to make.

However, when the two brains compete, there is either an emotional short-circuit—whereby the Old Brain may flood emotions, interrupting activities in the New Brain—or a stifling of the Old Brain when the New Brain squelches emotions.

Beneficial Emotional Short-Circuit

The Old Brain, also known as the Emotional Brain, is constantly on guard. When it perceives a danger or an exceptional opportunity, it interrupts activities in the New Brain.

Today, scientists have proven that when survival is at stake, the responses of the Old Brain are better at responding than are the abstract reflections of the New Brain because they are faster.

Example: Céline and Laura were deeply involved in conversation while waiting on the sidewalk for the traffic light to change. When the signal indicated that it was okay to "walk," Céline stepped off the curb. Laura immediately pulled her back. A car was illegally crossing the intersection and would have hit Céline. She was so close to it that the zipper of her jacket scraped the side of the car.

Laura's Old Brain perceived the danger. In an instant it interrupted the conversation—the activity of the New Brain—to enable the whole brain to act with its full resources until the danger had passed.

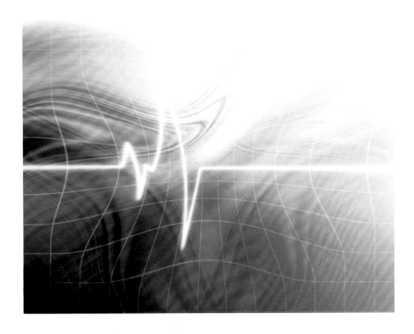

Harmful Emotional Short-Circuit

When your emotions are too strong, the prefrontal cortex of the New Brain no longer responds and loses its capacity to control behavior.

ABANDONED TO THEMSELVES, YOUR EMOTIONS DEPRIVE YOU OF REASONING ABILITY AND YOU MAY OVERLOOK IMMINENT DANGER.

When the Old Brain takes over all of the body's functions, the heart beats too fast, the stomach tightens, the digestion shuts down, the legs and the hands tremble, the whole body breaks out in a sweat.

Sometimes the body freezes and a flood of adrenaline knocks out the New Brain's functions. In that case, Laura would not have stopped Céline, who would have been hit by the car.

DO YOU KNOW?

People who experience PTSD (post-traumatic stress disorder) and anxiety attacks are no longer able to organize a coherent response to various life situations. Instead they are emotionally triggered by the association of a memory, which stimulates the Old Brain to react in a survival modality, even when there is no danger.

Example: Howard, an attorney, was an Army captain in the Vietnam War. His division was moving through the jungle and he was bringing up the rear. It was a beautiful, sunny day. A slight breeze was blowing through the palm trees with an aroma of jasmine in the air. Suddenly, at the next clearing, he saw all of his men blown to bits.

Fifteen years later, Howard was on a golf course in Miami. It was a sunny day, and a soft breeze was wafting through the palm trees with a fragrance of jasmine in the air. Howard hit the ground screaming and yelling in total panic and terror.

Smothering of the New Brain

Beneficial smothering of the New Brain
The New Brain has the ability to temper emotional reactions before they get out of hand. It frees you from a life controlled by instincts and reflexes only.

Harmful smothering of the New Brain
Excessive control of emotions fosters a temperament that lacks enough sensitivity to make appropriate decisions.

IF YOU STIFLE NATURAL EMOTIONS, YOU MAY EXPERIENCE DAMAGE TO YOUR HEALTH, BECAUSE THE OLD BRAIN DOES NOT HAVE THE PERMISSION TO INTERACT WITH THE BODY.

If you ignore distress, this distress expresses itself physically. You may experience symptoms such as:

- Unexplained fatigue
- High blood pressure
- Chronic colds
- Heart disease
- High cholesterol
- Intestinal disorders
- Skin problems

A lack of awareness of your own emotions brings a lack of awareness for others' emotions, as well as a feeling of loneliness. You may become unable to experience empathy or compassion for yourself or for others.

However, when the Old Brain and the New Brain cooperate, you are more comfortable with yourself, and thus can enjoy a sense of social ease and caring for others.

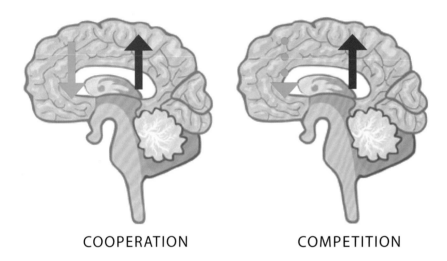

COOPERATION   COMPETITION

COMPETITION between the Old Brain and the New Brain makes you UNHAPPY.

COOPERATION between the Old Brain and the New Brain makes you HAPPY.

Dotted line = less neural flow
Solid line = more neural flow
Blue = new-brain
Red = old-brain

Harmful Stifling by the New Brain

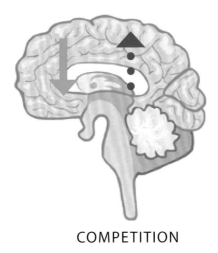

COMPETITION

Harmful Smothering of the New Brain

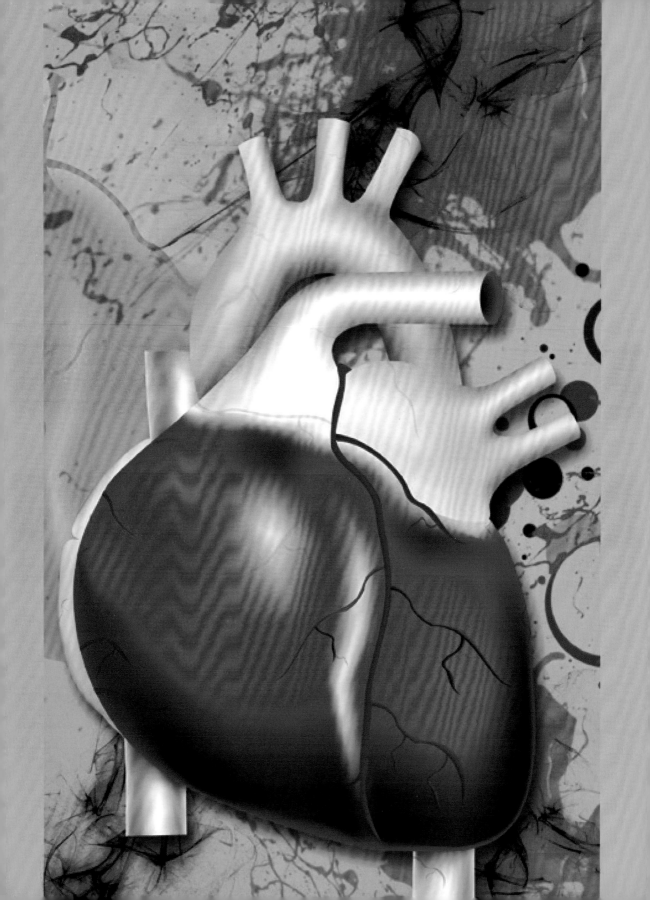

Chapter 4

The Importance of the Heart

Emotions are things that happen to us rather than things we will to occur. Although people set up situations to modulate their emotions all the time—going to movies and amusement parks, having a tasty meal, consuming alcohol or other recreational drugs—in these situations, external events are simply arranged so that the stimuli that automatically trigger emotions will be present. We have little direct control over our emotional reactions. Anyone who has tried to fake an emotion, or who has been the recipient of a faked one, knows all too well the futility of the attempt. While conscious control over emotions is weak, emotions can flood consciousness.
—**Joseph LeDoux,** *The Emotional Brain*[1]

THE EMOTIONS CAN be brought into harmony with the New Brain through attention and understanding of the human heart as a complex, organized and intelligent system that interacts with the two brains in the head.

It was first thought that all emotions and sensations originated externally and were picked up and organized by the New Brain.

It was then discovered that most information came not from the outside, but from the Old Brain through the thalamus, which is a bridge between the Old Brain and the New

Brain. **TODAY IT IS THOUGHT THAT THE HEART SENDS ITS MESSAGES UP THE SPINE, THROUGH THE VAGUS NERVE, AND ALLOWS THE NEW BRAIN TO UNDERSTAND WHAT THE BODY IS DOING.**

The thalamus is divided into two lobes with long, nerve fibers extending into all cortical areas. The thalamus is heavily involved in sensory processing and it is the thalamus that makes the decision whether or not you react in a mostly emotional manner or in a mostly cognitive manner to any life situation.[2]

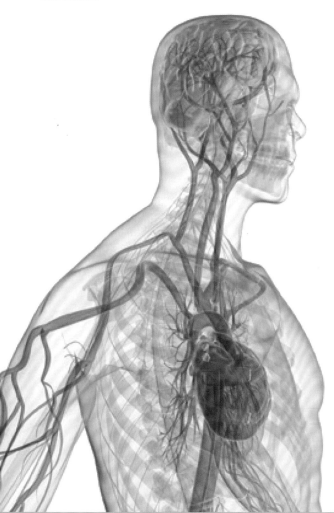

THE THALAMUS BELIEVES THAT THE MORE EMOTIONALLY CHARGED YOUR BODY IS IN REGARD TO AN ISSUE, THE MORE LIFE-THREATENING A SITUATION ACTUALLY IS.

If this is not the case in reality, your entire emotional being starts living a lie. Your conscious, thinking New Brain picks up on the fact that your reactions are untrue. In most cases it will do the opposite of what needs to happen to

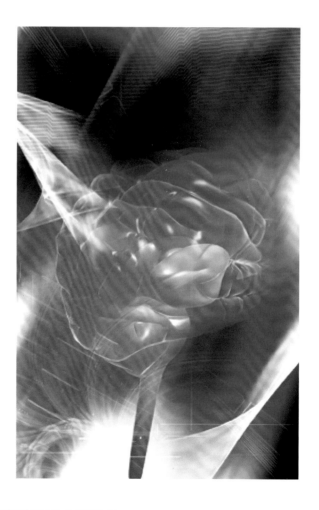

release the untruth. It will hold on to feelings and not allow their release from the body.

The Heart Brain can intercede in this negative process and send truthful messages to the Old Brain, which can be transferred over the thalamus into the New Brain. The New Brain will then release its hold on the negative emotions.

The heart sends its messages up the spine through the vagus nerve and allows the New Brain to understand what the body is doing. The heart balances and supports the ongoing systems in the body, including the lungs, stomach, intestines, voice box and the ears. **THE HEART ALWAYS REPORTS THE TRUTH OF WHAT THE STATE OF THE BODY IS TO THE NEW BRAIN.** Being in touch and in harmony with your heart allows true messages about all your organs' systems to be known by your New Brain. This communication then allows your New Brain to act accordingly to order balance, harmony and homeostasis throughout the body.

The Heart-Math Institute in Boulder Creek, California is dedicated to the study of the intelligence of the heart. Many systems and techniques have been developed through Heart-Math research to facilitate awareness of the importance of the harmony of the heart and of the emotions.[3]

Today, we know that most information received by the Old Brain comes from the organs. The heart gathers most of that information from those organs and sends it to the Old Brain.

DO YOU KNOW?

The reason you can't change your emotional responses using your thinking is because all of your senses are routed to enter *behind* your thinking in your Old Brain. Your visual center, for example, is at the rear of your brain, and visual signals entering your brain do so *below* your New Brain in the Old Brain.

Example: When you see something, you think you see it because it is in front of your eyes. Wrong! It is the heart which communicates information to your Old Brain up through the vagus nerve.

Sensation and Feeling Begin in the Heart

What is more surprising is that the Old Brain works like a filter to the New Brain. It may or may not allow emotional information to be passed to the New Brain. As you have just learned, the thalamus is the bridge between the Old Brain and New Brain and, depending on the information, may be blocked or opened by the Old Brain.

Body information > Heart > Old Brain > New Brain

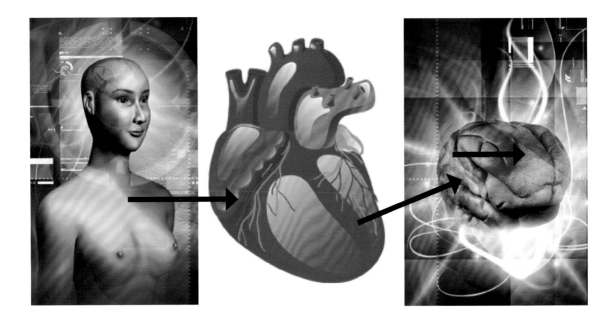

When the sensations or feelings that the heart sends to the brain in the head are too disturbing, the Old Brain accelerates or slows down the beating of the heart. Sometimes the Old Brain is flooded with emotions. You may have too much stress or excitement. At that time, your heart may beat too rapidly and your breathing may be interrupted. To remedy this situation, the heart must be contacted. Breathing in rhythm must begin.

In the past, the Old Brain was considered useful for only lower primates and reptiles to house instincts. The New Brain was given credit for being able to see and understand the more highly-developed sensations and emotions in the human being.

Today, we know this is not true. **WE UNDERSTAND THAT THE OLD BRAIN IS NOT TO BE OVERCOME OR SUBDUED**. It is a valuable, huge warehouse of information which is gathered from the various organ systems of the body.

We see the "Emotional Brain" (Old Brain) as a command post that continually receives information from different parts of the body.[3]

The Heart Brain gathers information and sends it to the Old Brain, which allows the New Brain to weigh and discern the best course of action for the entire person.

This is the key to the healthy functioning of the entire being. Scientists talk of a "Heart Brain System."

IF THE HEART IS NOT IN HARMONY—RACING, PANICKED, FRIGHTENED OR STRESSED—IT WILL SHUT DOWN ALL BODILY INFORMATION EXCEPT FOR CERTAIN VITAL FUNCTIONS.

Indeed, when the heart is stressed, digestion comes close to a standstill. The adrenal glands are called to attention waiting to fight or flee. The amount of blood circulating comes out of the head and into the muscles and nerves to strengthen the body to face the fearful task at hand.

During this kind of situation, the Old Brain may know very little about what is truly going on in the body and therefore may transfer even less information to the New Brain for processing. Many people report that stress leaves them checked out, unconscious, and in a state with no ability to think at all. Any action is therefore guided from an instinct to survive, even if in reality there is no survival threat at hand.

It is, therefore, imperative that we come back to the heart, that we **ALLOW THE HEART TO FUNCTION IN A KIND OF PEACEFUL HARMONY, SO THAT IT IS WILLING TO RELAY THE BODILY INFORMATION TO THE OLD BRAIN IN THE HEAD.** This heart harmony, in turn, allows a free flow of that information to the New Brain. The New Brain is able to cognitively come to some understanding of where and what is happening in the total body organism and make appropriate judgments for the wellbeing of the whole person.

Why is the heart so important?

It's not just a pump! And it is not just the strongest muscle in the body. It is so much more than that. **WHEN THE HEART IS IN HARMONY, IT ALLOWS FOR A HOMEOSTATIC CONDITION IN THE WHOLE HEART BRAIN SYSTEM,** and

harmony among all physical functions in the body, so that you can be present and aware of your physical, mental and emotional states as well as their relationships to your existing environment. Further, this heart harmony allows you to move, speak and act in ways that will bring more wellbeing and peace.

Example: If your toe is severely infected, the heart will relay the information to the Old Brain, and then the Old Brain will relay it to the New Brain and the New Brain may perceive that maybe you need to go to the doctor and do something about the infection in your toe. How many times have you heard the story of someone who supposedly had no symptoms of anything wrong and then by the time that person got to the doctor, it was too late to treat the illness because it spread too far?

By listening to your heart and by doing the exercises in chapter 11, a new sense of awareness can grow within you. You will not only know and understand the various functions of your body, but you will become keenly aware of the difference between being in balance and being out of balance. The Heart Harmony Exercises will allow you to become more aware of what is going on inside your body, so that you may use your Heart Brain system to avert any kind of serious, chronic or severe illness.

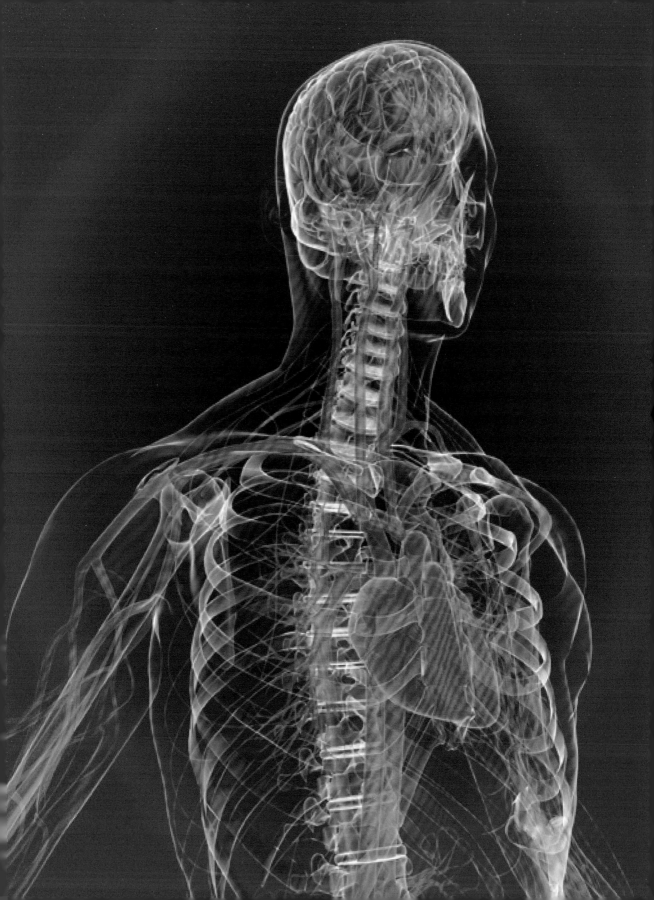

CHAPTER 5

The Brain in the Heart

As you learned in previous chapters, the Heart Brain gathers information from all the bodily systems: neural, cardiovascular, hormonal and immune.

- ♥ It adapts its behavior according to its perceptions.
- ♥ It creates its own memories and uses its experiences to choose its responses.
- ♥ It has a small hormone factory.

The Brain in the Heart

- ♥ Produces its own adrenaline.
- ♥ Releases ANF (atrial natriuretic factor) which regulates blood pressure.
- ♥ Secretes oxytocin (used when mother breast feeds, during courtship, intercourse, etc.).
- ♥ Has a close connection with the brain in the head through the vagus nerve and the nerves in the spinal column.

How the Brain in the Heart Works

The decisions made within the heart directly impact the way the two brains in the head perceive and process information. The signals sent by the Heart Brain escalate into the Old Brain then into the New Brain where they influence perception, decision making and other cognitive processes.

- ♥ In states of stress—anxiety, fear, anger, worry, sadness—your heartbeat is irregular or "chaotic."

- ♥ In states of wellbeing—joy, compassion, love, gratitude—the heartbeat is regular, or "in harmony."

States of Stress = CHAOS

- ♥ Anxiety
- ♥ Fear
- ♥ Anger
- ♥ Worries, sadness

States of Wellbeing = HARMONY

- ♥ Joy
- ♥ Compassion
- ♥ Love
- ♥ Gratitude

The Heart-Old Brain System

Scientists have discovered that when the Old Brain does not function at its best, not only does this stress the heart, but it also stresses the New Brain. This system of Heart-Old Brain-New Brain is now referred to in science as the Heart Brain System.[1]

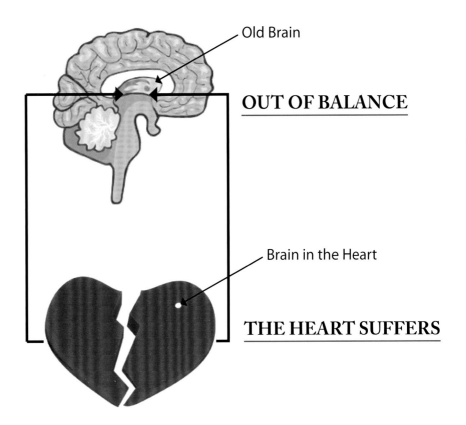

The relationship between the Old Brain and the brain in the Heart is key to understanding balance/imbalance in the Old Brain.

WHEN THE OLD BRAIN IS OUT OF BALANCE, THE BRAKE DOES NOT FUNCTION ANY MORE, AND THE HEART SUFFERS AND BECOMES EXHAUSTED.

There are two branches of nerves that travel along the spine to the head. These branches are named "accelerator" and "brake." They connect the heart to the Old Brain. These branches need to be equally strong to counterbalance each other and to avoid one being dominant over the other.

The accelerator and the brake, scientifically named "autonomic nervous system" are constantly speeding up and slowing down the beats of the heart. This constant change in the heartbeat is why the interval between two successive heartbeats is never identical. This rate variability is healthy.

AN UPSET HEART IS OUT OF HARMONY. IT IS IN CHAOS AND THE HEART BRAIN SYSTEM DOES NOT FUNCTION WELL.

You go into a mode of "fight or flight." In this mode the heart races and this becomes dangerous. Why is it dangerous?

The hormones normally secreted by the heart, such as adrenaline, estrogen, progestin, become unavailable to the brain. The heart sends little information about what is going on in the body to the Old Brain. As a consequence, the heart loses its ability to monitor the homeostatic conditions in the body.

THIS DISCONNECTION THEN ALLOWS FOR THE GROWTH OF VIRUSES, BACTERIA, FUNGI, AND OTHER INVADERS.

Sometimes a person suffers from tachycardia (violent accelerations of the heartbeat lasting several minutes) or from arrhythmia (abnormal heart rhythm), or from anxiety attacks. This lack of connection often leads to a very serious condition, which has gone undetected by the New Brain, and

therefore, has been allowed to develop to either the extreme detriment to the person or even death from conditions such as, cancer, stroke or chronic autoimmune disease.

It is not until the heart comes back into a state of harmony that it is willing to reconnect with the Old Brain, and then the Old Brain is willing and able to share information with the New Brain.

BUT WHEN THE OLD BRAIN IS OUT OF BALANCE, THE BRAKE DOES NOT FUNCTION ANYMORE AND THE HEART SUFFERS AND BECOMES EXHAUSTED.

As you know, the heart is closely connected with the head. Together, they are a true Heart Brain System. Within this system, the three organs (Heart Brain, Old Brain and New Brain) constantly influence each other.

The autonomic nervous system connects the brain to the heart. Within this system are two branches of nerves:

Sympathetic branch (also called the "accelerator branch") speeds up the heart and activates the Old Brain by releasing adrenaline and noradrenaline.

*Parasympathetic branch (*also called the "brake branch"), acts as a brake on the heart and on the brain by releasing acetylcholine that promotes states of relaxation and calm.

These two branches of nerves regulate the functioning of your organs and particularly the heart.

Input from the Old Brain to the Heart

THE BRAKE AND ACCELERATOR ARE CONSTANTLY SPEEDING UP AND SLOWING DOWN THE HEART BEAT. To understand this phenomenon, visualize a scale with two equal platforms that are continuously adjusting to keep balance. The consequence of the back and forth movement, like the scale, is the normal variability between consecutive heartbeats.

When there is chaos in your cardiac rhythm, emotions may seem overwhelming, because there is chaos in the physiology of your body.

These emotions can be so strong that the New Brain is unable to exert its judgment and discernment allowing the reflexes and instincts to take over. You may become extremely emotional and irrational. An example is a severe panic attack.

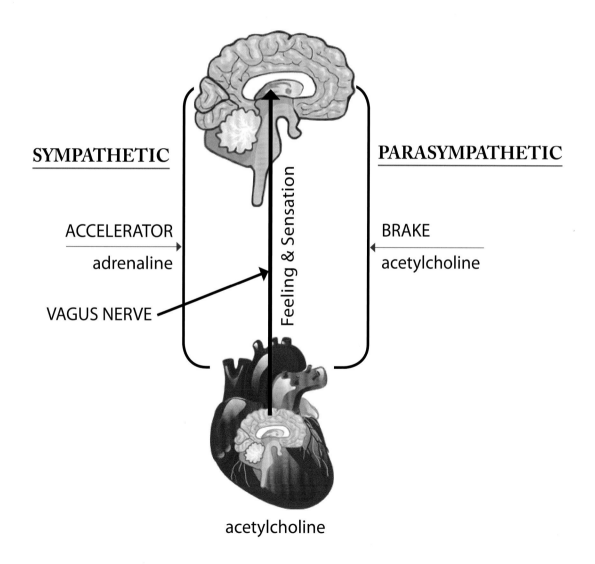

Input from the Heart to the Brain in the Head

The vagus nerve, which is part of the parasympathetic nervous system, is the highway on which feelings and sensations travel back and forth between the Heart Brain and the Old Brain. These sensations affect the functioning of every organ and influence the blood flow through all vessels of the body. One of the functions of the heart is to release hormones and regulate blood pressure.

THE SIGNALS FROM THE HEART BRAIN ESCALATE INTO THE OLD BRAIN THEN INTO THE NEW BRAIN, WHERE THEY INFLUENCE PERCEPTION, DECISION-MAKING, AND OTHER COGNITIVE PROCESSES.

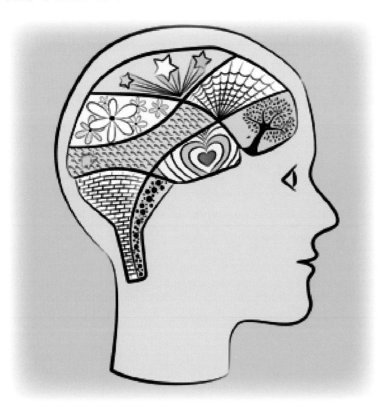

DO YOU KNOW?

During the time of courtship, when your heart is thumping or when you blush, it is because the accelerator branch speeds up the heart. If you want to calm down, just take a deep breath and you will recover your senses. You have then pressed on the brake.

CHAPTER 6

Emotional Mastery

The relationship between the Old Brain and the Heart Brain is one of the keys to emotional mastery.

As you know, when the heart comes into a state of harmony there is a free flow of information from the Heart Brain to the Old Brain to the New Brain and then back again to the heart and the Heart Brain. The balance of adrenaline and noradrenaline is maintained.

Vital organs, such as the kidneys and liver, are not over stressed and can make appropriate amounts of hormones. They can be directed to other important organs, such as the pancreas and the spleen. The pancreas can receive insulin necessary to monitor blood sugar. The spleen then receives new red blood cells that can be released into the body when the old ones have dissipated.

EMOTIONALLY, THE BALANCE IN THE HEART BRAIN SYSTEM IS VITAL TO ALLOW A SENSE OF WELLBEING. Scientists tell us that sleep problems are rampant today because of the lack of good feelings in the heart, and the heart's resistance to sending appropriate hormones to the brain. Some of the hormones that exist in the brain are serotonin and melatonin. Low levels of these hormones make it difficult to fall asleep and stay asleep through the night.

FURTHER, THE NUMBER ONE EMOTIONAL DISEASE IN THE UNITED STATES IS DEPRESSION. Depression comes from low levels of serotonin in the brain. Both sleep disorders and depression could be remedied and emotional balance restored through the communication, harmony and coherence of the entire Heart Brain System.

DO YOU KNOW?
Many travelers use melatonin to fight jet lag when they go over numerous time zones.

Please, do not forget that the heart actually perceives and feels. It is not the brain that does this. Both brains in the head interpret what the heart perceives and feels. As stated previously, when you describe emotions, you use the word "heart."

WHEN THE OLD BRAIN IS OUT OF ORDER, THE HEART SUFFERS AND WEARS OUT.

Example: John

John, a healthy fifty-year-old college professor, had a medical checkup. All results came back favorable. A few days after this checkup, John's only child, a 21-year-old daughter, died suddenly of a rupture of the hepatic vein. Six months later, John suffered a heart attack. The doctors saved him, but he was inconsolable and depressed because of his daughter's death. Soon after, John had a second heart attack and died.

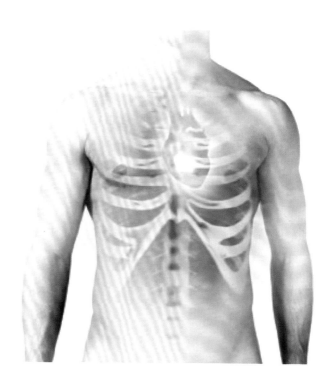

TODAY, CARDIOLOGISTS AND PSYCHIATRISTS KNOW THAT STRESS IS AN EVEN GREATER RISK FACTOR FOR HEART DISEASE THAN SMOKING.

Doctors know that there is no drug to harmonize the relationship between the heart and the brain.

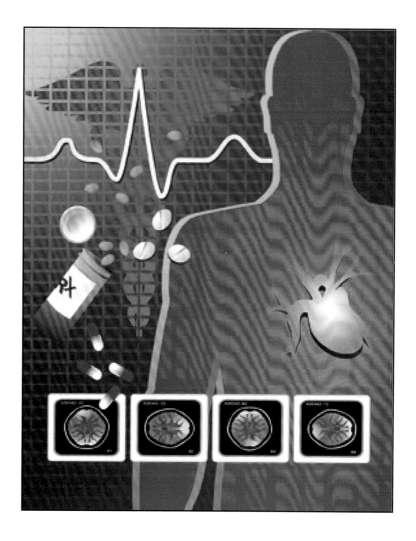

By utilizing the exercises in Chapter 11, you will learn how to create this harmony almost instantly. John, the heart broken patient, could be alive today. Unfortunately, his doctor did not know about these exercises.

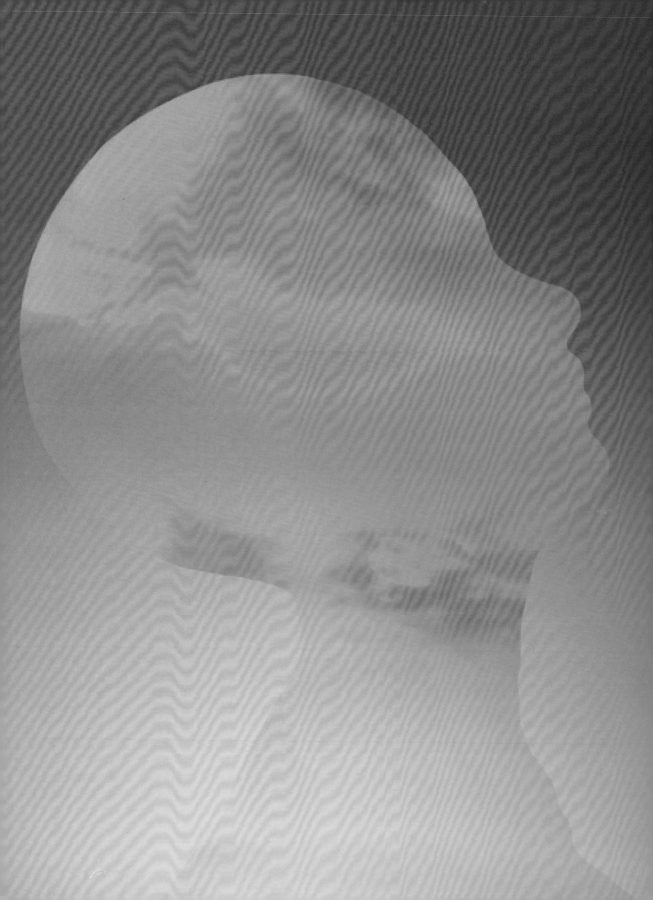

CHAPTER 7

Everything Starts with Emotions

Your heart is always communicating with your body, and as a consequence, also communicating with your Old Brain.

Your heart communicates through

- ♥ Neurological pathways
- ♥ Hormonal pathways
- ♥ Other pathways

Through your heart you realize that your intuitive feelings are right because they bring you a sense of peace or ease. You also realize that your judgmental feelings bring you negative reactions, such as irritation, hurt and anger. Your positive feelings enable you to approach difficulties with a proactive and broad perspective.

Bodily systems such as digestion, blood pressure and respiration synchronize with cardiac rhythm. When harmony exists in the heart, you have more mental clarity, creativity, intuition and better problem solving abilities.

As you know, **YOUR OLD BRAIN—THE ONE YOU SHARE WITH MAMMALS—CARRIES YOUR DESIRES AND IS THE ONE WHICH GIVES YOU PASSION**. In order to bring that passion out, you call for the help of your New Brain. The New Brain supports passion intellectually through its knowledge, ability to imagine, ability to concentrate and ability to see the future, so that you can attain those goals for which you have passion.

YOUR INTUITION ALLOWS YOU TO ACTIVATE YOUR IMAGINATION, WHICH PROVIDES YOU WITH MANY MORE CHOICES. These choices can be utilized in many situations to resolve life's problems and to reduce stress. As resolution occurs, the New Brain, the Old Brain and the Heart Brain are functionally and optimally in harmony and balance.

DO YOU KNOW?

Stress, even small, unidentified stress, affects health. Too much stress contributes to and agitates many health problems such as heart disease, high blood pressure, stroke, depression and sleep disorders.

CHAPTER 8

The Electromagnetic Fields of the Heart

ELECTROMAGNETIC FIELDS (EMFs) are physical fields produced by electrically charged atoms. Moving electrical currents produce magnetic fields. These electromagnetic fields are natural forces in the universe which exist in space and time around all living things. EMFs are present in everyone and everything in the known and unknown environment. According to James Oschman, "The biomagnetic field of the heart extends indefinitely into space. Even though its strength diminishes with distance, there is no point at which we can say the field ends."[1]

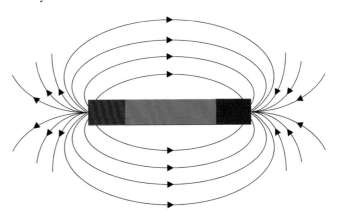

EMFs either pass through the heart without interaction or they interact directly with the heart. The heart muscle itself, because of its electrical activity, creates its own EMFs.

Over the past two hundred years, the focus of study for electromagnetic fields has been the brain. **ALTHOUGH SCIENTISTS KNEW THAT THE BRAIN PRODUCES ELECTROMAGNETIC FIELDS, IT IS ONLY WITHIN THE LAST FORTY YEARS THAT THEY HAVE TURNED THEIR ATTENTION TO THE ELECTROMAGNETIC FIELDS PRODUCED BY THE HEART.** Scientific progress in instrumentation has revealed the intense strength of the heart's magnetic fields, which have been shown to be the most powerful of all the magnetic fields in the human body.

DO YOU KNOW?

Research shows that a specific breathing technique can enhance cardio-electromagnetic communication. This exercise is performed by inhaling to the count of four, immediately exhaling to the count of six, and pausing to the count of two. By repeating this eight times and combining it with a feeling of great appreciation for life, the heart's magnetic fields have been found to significantly increase and positively affect the cardiac functions—heart rate, variability, blood pressure, and respiration.[2]

In 1925 Florence Scovel Schinn published a metaphysical book, *The Game of Life and How to Play It*. Her philosophy centered on the power of positive thoughts and actions. Schinn's advice was generally accompanied by a real life anecdote: An upset client came to Schinn complaining about her employer. The woman said, "My employer is cold and critical and seems to want to fire me, and I have done nothing wrong." Schinn asked the client to visualize her employer while breathing in and feeling her heart, and sending her employer love instead of anger. Ready to try anything, the unhappy woman said, "Very well, I'll do it." A week later, the astonished client told Schinn, "I've been practicing sending love to my employer, and suddenly she is kind to me." Schinn knew intuitively that **POSITIVE THOUGHTS AND FEELINGS ARE POWERFUL, MAGNETIC FORCES IN THE UNIVERSE**. Sending love increases the electromagnetic fields of other people, allowing them to receive love.[3]

In the past the lack of modern technological devices hampered science's ability to accurately measure the power of the heart's electromagnetic fields. Only recently have scientists been able to verify that the heart's EMFs are stronger than those of the brain. **THE ELECTRICAL IMPULSES OF THE HEART ARE AT LEAST ONE HUNDRED TIMES STRONGER THAN THOSE OF THE BRAIN,** and the magnetic fields of the heart are at least five thousand times stronger than those of the brain.

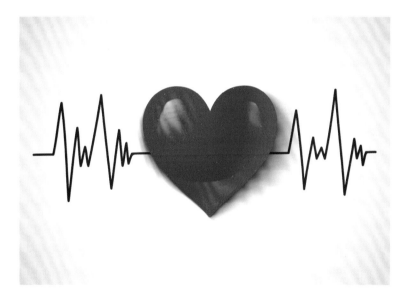

With these new understandings research began to illustrate how electromagnetic fields emanating from one's heart affect other people, animals, plants, and the environment. What was once thought of as only a blood pumping mechanism for the human body, the heart is now seen as a complex organism that transports concentrated heart emotion, not only to loved ones nearby, but to others as far away as the other side of the earth and beyond.

According to Dr. Gregg Braden, quantum physics has proven that a human being's presence in a field of energy actually changes the nature of that field.[4]

MORE RECENTLY PHYSICISTS HAVE PROPOSED THAT INTENTIONAL THOUGHTS THROUGH THE HEART AFFECT THE ELECTRICAL BALANCE OF ATOMS, and the very shape of an atom begins to change as well as its nature. Furthermore, it has been proven that water retains changes induced into it by exposure to a strong, concentrated electromagnetic field, as demonstrated in the next paragraph.

There is evidence that suggests that intentional, concentrated feelings of love, gratitude, and appreciation actually change the nature of drinking water. Dr. Masaru Emoto, doctor of alternative medicine in Japan, found that water in its frozen state reveals the power of electromagnetic fields, and that water is deeply connected to individual and collective consciousness. In an experiment Dr. Emoto filled several bottles with tap water. Each bottle was wrapped with  a written note that conveyed either a positive or a negative message. The writing was taped on the bottle, facing inside. When the writing said, "Thank you," and the water was frozen, the crystals were complete and lovely. Conversely, when the writing had a negative message, "You Fool," the frozen crystals became incomplete and malformed.

KNOWING THAT THE ADULT HUMAN BODY IS MORE THAN 70% WATER AND AN INFANT'S BODY IS MORE THAN 90% WATER, HUMANS CAN BE HURT PHYSICALLY AND EMOTIONALLY JUST LIKE THE WATER.[5]

> THE FOCUSED HUMAN EMOTION THAT WAS EXPRESSED ALL AROUND THE WORLD SENT ELECTROMAGNETIC PULSES BEYOND THE EARTH INTO OUTER SPACE.

Complex satellites called Geosynchronous Environmental Satellites (GOES) continuously monitor the electromagnetic fields of the earth. Two GOES keep watch over the nature and changes of the electromagnetic fields around the earth—one in North America and one in South America. During 2001, a huge change in the fields was recorded and found to have begun on September 11, just after the first plane hit one of the Twin Towers in New York. The focused human emotion that was expressed all around the world sent electromagnetic pulses beyond the earth into outer space. This finding is phenomenal and illustrative of how powerful focused human emotion can be. Noting this effect of focused human emotion on the earth's electromagnetic fields challenges old scientific beliefs about the function and purpose of the human heart.[6]

Scientific instruments have shown that when two people are in love each of their electromagnetic fields extends to the other, creating a beautiful heart-shaped field between the two.[7]

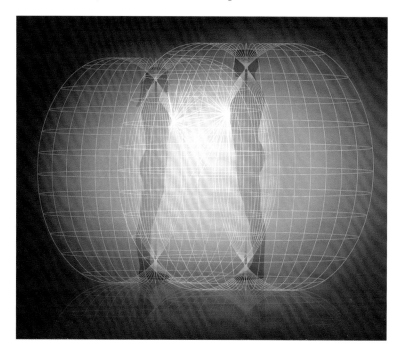

Furthermore, recent research suggests that the heart may play a significant role in healing disease. As the heart sends out a powerful, electromagnetic field it actually affects and changes the pathology of patterns in the sick person to the extent that heartfelt emotion in the form of EMFs can ameliorate and resolve illness and distress.

NOT ONLY DO HEARTFELT EMOTIONS CHANGE PATHOLOGICAL PATTERNS, THEY ALSO CARRY INFORMATION THAT CAN INTERACT WITH ORGANS AND OTHER STRUCTURES IN THE BODY TO STIMULATE DYNAMIC ACTIVITY.

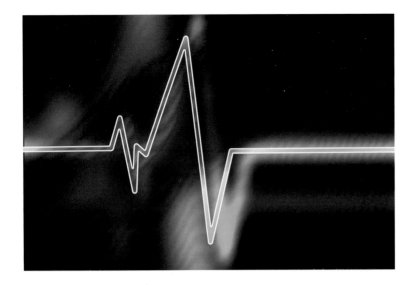

Research at the Heartmath Institute teaches us that information pertaining to a person's emotional state is communicated throughout the body via the heart's electromagnetic field. More specifically, the Institute has demonstrated that sustained, positive emotions seem to allow for an elevated psycho-physiological function, increasing efficiency and harmony in the activity and interaction of the body's systems. At a psychological level, this is linked to a reduction in internal mental dialog, and perceptions of stress. It is also linked to a sense of increased emotional balance, mental clarity, intuitive discernment and cognitive performance.[8]

> MORE SPECIFICALLY, THE INSTITUTE HAS DEMONSTRATED THAT SUSTAINED, POSITIVE EMOTIONS SEEM TO ALLOW FOR AN ELEVATED PSYCHO-PHYSIOLOGICAL FUNCTION, INCREASING EFFICIENCY AND HARMONY IN THE ACTIVITY AND INTERACTION OF THE BODY'S SYSTEMS.

New data suggests that intuition could not exist without the heart's EMFs. Energetic information that exists in the electromagnetic fields goes beyond space and time. Extensive research reveals evidence that both the heart and the brain receive and respond to information about a future event before the event actually occurs. The heart appears to receive intuitive information before the brain. This may suggest that the heart's fields are linked to a greater, subtle, energetic field containing information about objects and events remote in space and ahead in time. Karl Pibram has called this field the *spectral domain*. This is a fundamental order of potential energy that encompasses both space and time and is thought to be the basis of our consciousness.[9]

DO YOU KNOW?

Recent studies conducted by the HeartMath Institute explain the sense of wellbeing and peace that people experience in the presence of horses. The electromagnetic field generated by a horse's heart is five times larger than that of the human heart. The horse's EMFs have a spherical shape that completely envelop the human body. These electromagnetic fields are stronger than the human EMFs and can actually influence human heart rhythm. One can experience the benefits of clarity, peaceful existence, and joy by being in the presence of a horse.[10]

An Exercise

Here is an exercise that you can do with your heart to begin to understand the far-reaching effects it can have. Think of someone whom you love. For seven days take time each day to breathe and send love from your heart to that person. On the seventh day, call your loved one and ask how life has been going. Does he/she feel your love?

CHAPTER 9

The Language of the Heart

There are many expressions in the English language that relate to feelings that include the word "heart."

LANGUAGE EXPRESSES THE VALUES OF A CULTURE, AND IN THE ENGLISH LANGUAGE THE HEART IS ASSOCIATED NOT ONLY WITH FEELING, BUT WITH WISDOM, KNOWLEDGE, AND UNDERSTANDING.[1]

DO YOU KNOW?

Your eyes never lie. A forced smile, the one that is produced for social reasons, mobilizes only the muscles around the mouth, and you can see the teeth of the smiling person. A genuine smile adds a mobilization of the muscles around the eyes. The eye muscles do not contract at will. They respond to the control of the deep Old Brain, the Emotional Brain, and the heart.

Remember, your feelings start in your heart, not in your head.

The following expressions are examples of the importance of the "heart" to the human experience, and they exist for a reason. Even without having a scientific background, we humans have always innately known the importance of the heart and how it speaks to us.

- ♥ **Cheating heart.** A person with a cheating heart does not honor the vows of monogamy.

- ♥ **Learn by heart.** To learn by heart is to learn something word for word.

- ♥ **All heart.** A person who is all heart is very kind and generous.

- ♥ **Heart of gold.** Someone with a heart of gold is a genuinely kind and caring person.

- ♥ **Heart of glass**. A person with a heart of glass is easily affected emotionally.

- ♥ **Heart of steel**. Someone with a heart of steel does not show emotion or is not affected emotionally.

- ♥ **Heart isn't in it.** When one's heart is not in something, he or she does not believe in it or support it.

- ♥ **Heart in your mouth.** To have your heart in your mouth means that you feel nervous or scared.

- ♥ **Heart in your boots.** To have your heart in your boots means that you are very unhappy.

- ♥ **Heart misses a beat.** If your heart misses a beat, you are suddenly shocked or surprised.

- ♥ **Heart to heart.** A heart-to-heart is a frank and honest conversation with another, discussing issues honestly and plainly, no matter how painful.

- ♥ **Heart in the right place.** To have the heart in the right place means to be good and kind, in spite of appearances.

- ♥ **Have a heart.** A person with a "heart" is someone who is kind and sympathetic. Expressing "Have a heart" to someone is to ask for understanding and sympathy.

- ♥ **Melt your heart.** If something melts your heart, it affects you emotionally and you cannot control the feeling.

- ♥ **Bare your heart.** If you bare your heart to someone, you tell that person your personal and private feelings.

- ♥ **Change of heart.** If you change the way you think or feel about something, you have a change of heart.

- ♥ **Break your heart.** If someone upsets you greatly, he/she breaks your heart, especially if he/she ends a relationship.

- ♥ **My heart bleeds.** If your heart bleeds for someone, you feel genuine sympathy and sadness for him or her.

- ♥ **Bleeding heart.** A bleeding heart is a person who is excessively sympathetic towards other people.

- ♥ **Not have the heart.** If you don't have the heart to do something, you don't have the strength or courage to do something (usually used in the negative).

- ♥ **Close to your heart.** If something is close to your heart, you care a lot about it.

- ♥ **After your own heart.** A person after your own heart thinks the same way that you think.

- ♥ **My heart goes out to you.** If your heart goes out to someone, you feel genuine sympathy for him or her.

- ♥ **Queen of hearts.** A woman who is pre-eminent in her area is a Queen of Hearts.

- ♥ **Wear your heart on your sleeve.** To wear your heart on your sleeve is to show your emotions and feelings publicly.

- ♥ **Put your hand on your heart.** To place your hand on your heart means that you know it to be true.

- ♥ **Eat your heart out.** If someone tells you to eat your heart out, that person is saying that he or she is better at doing something than you are.

- ♥ **From the bottom of my heart.** When someone says "from the bottom of my heart" that person means that it is with emotion and great feeling.

- ♥ **Warm the cockles of your heart.** If something warms the cockles of your heart, it makes you feel happy.

- ♥ **Faint heart never won fair lady.** This means that you will not get the partner of your dreams if you lack the confidence to let them know how you feel.

- ♥ **Hale and hearty**. Someone who is hale and hearty is in very good health

- ♥ **Tug at the heart strings**. When something tugs at your heart strings; it makes you feel sad or sympathetic towards it.

- ♥ **In a heartbeat.** If something happens very quickly or immediately, it happens in a heartbeat.

- ♥ **Absence makes the heart grow fonder.** This means that when people are apart, their love grows stronger.

You are invited to choose one expression from the list that you particularly relate to and copy it on this page or in any special place. It may serve as a gateway to meet again with your inner self if you have lost it.

_____
_____
_____
_____
_____
_____
_____
_____
_____
_____
_____
_____
_____
_____
_____
_____
_____
_____
_____
_____
_____
_____
_____
_____

SOME OF YOU, OVER THE YEARS, HAVE LEARNED TO IGNORE YOUR INNER SELF IN ORDER TO COPE WITH LIFE'S MANY RESPONSIBILITIES AND DISAPPOINTMENTS. You have had to deal with your boss, take care of your children, find time to visit your parents, etc. Know consciously that you may have stifled your feelings of frustration. You may have felt deprived when you did not have the time to play golf, jog, go to the theater, walk on the beach, stay in bed on Sunday morning, etc., and you became stressed.

Chapter 10

Heart Harmony

When the heart is in harmony, your body's immune systems function more optimally, thereby assisting you in the recovery process. However, when you have an ongoing illness, you feel a loss of control. The Heart Harmony Exercise in Chapter 11 gives you back some control over your illness. Practicing this exercise gives you the ability to reach some self-management of your pain and stress.[1]

The exercise is a safe method of self-management, which does not require medication. However, if you are on medication this exercise can be an added bonus.

YOU CAN INTENTIONALLY CHANGE THE PATTERN OF YOUR HEART RHYTHM INTO A SMOOTH ONE, AND HELP THE NERVOUS SYSTEM AND THE BODY IN GENERAL. YOUR HEART RHYTHM CHANGES AS YOU PRACTICE EACH EXERCISE.

The Heart Harmony Exercise helps reduce stress and attain emotional self-regulation and self-control. In addition to relief from chronic illnesses, you will also receive benefits for psycho-physiological conditions.

Here is a Simple Explanation:

Your heart rate is sensitive to different events taking place in your mind and body.

MOST PEOPLE THINK THAT THE HEART BEATS AT A VERY STEADY, UNCHANGING RATE ALL DAY, BUT IT DOESN'T. YOUR HEART BEATS FASTER AND SLOWER FOR ALL KINDS OF REASONS.

When you climb stairs the muscles in your body need more blood, nutrients and oxygen—thus your heart rate naturally increases. When you sleep your heart rate decreases, because the muscles in your body are at rest.

Remember, there is no drug to harmonize the relationship between the heart and the brain; however, the Heart Harmony Exercise is an excellent tool to achieve this goal.[2]

The Heart Harmony Exercise Allows You To:

- ♥ Control anxiety and depression
- ♥ Lower blood pressure
- ♥ Increase the hormone DHEA
- ♥ Stimulate the immune system
- ♥ Turn back the physiological clock
- ♥ Clear your mind
- ♥ Find your profound desires
- ♥ Make existing knowledge more readily accessible when stressed
- ♥ Deal with insomnia
- ♥ Reduce stress
- ♥ Eliminate panic attacks
- ♥ Reduce depression[3]

DO YOU KNOW?

Cardiologists and psychiatrists have brought to light that depression coming within six months after a myocardial infarction will more accurately predict a death for the patient than most measurements of the heart function.[4]

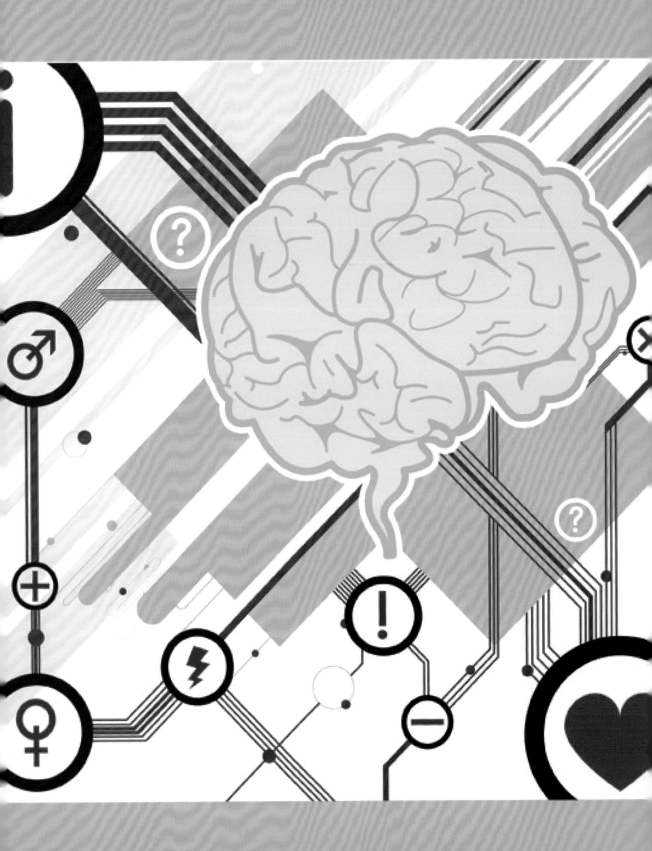

CHAPTER 11

Heart Harmony Quiz

What is Stress?

STRESS IS USUALLY defined as a state of negative emotional arousal, which is associated with feelings of discomfort or anxiety.[1] Stress is attributed to outside, unpleasant events.

STRESS AFFECTS YOU PHYSICALLY, MENTALLY AND EMOTIONALLY.

Scientists have demonstrated that it is not unpleasant external events that stress you. It is your emotional reactions to these events.

At the University of Pennsylvania, in the laboratory of Martin Seligman, Ph.D., it was demonstrated that it is the perception of helplessness you may have in the face of stress that affects your body and not the stress itself.[2]

To find out the areas in your life where you experience stress, ask yourself the following questions:

I am anxious about:

- ❏ Getting to the airport on time
- ❏ Being sick
- ❏ Locating a parking place
- ❏ Taking a challenging test
- ❏ Confronting my teenager
- ❏ Losing my job
- ❏ Getting enough sleep
- ❏ _____
- ❏ _____

I have repeated:

- ❏ Headaches
- ❏ Backaches
- ❏ Colds
- ❏ Sinus infections
- ❏ _____
- ❏ _____

I always worry about:

- ❏ Money
- ❏ War
- ❏ Job Security
- ❏ Family relationships
- ❏ Friends
- ❏ _____
- ❏ _____

I am angry when I have to deal with:

- ❏ Spouse
- ❏ Children
- ❏ My mother
- ❏ Household chores
- ❏ Paying bills
- ❏ Mother-in-law (father-in-law)
- ❏ _____
- ❏ _____

If you have checked 3 boxes or more, then you may have unknown stress.

How Does Stress Affect You?

At the end of 2010, scientists at Stanford University declared that 90% of all doctor visits were primarily related to stress. Eight out of 10 medications (antidepressants, medication for high blood pressure, etc.) prescribed in the United States are for diseases related to stress.[3]

Stress affects you:

- ❏ Physically
- ❏ Mentally
- ❏ Emotionally

DO YOU KNOW?

Scientists have found that unmanaged emotional stress is as much as six times more predictive of cancer and heart disease than cigarette smoking, cholesterol or blood pressure.[4]

Chapter 12

Heart Harmony Exercises

The basic science of mind-body unity suggests that every shade of emotion and every facet of selfhood is curled together with body states relevant to health.
—Henry Dreher[1]

Fanny was having a full body massage. When the masseuse worked on her temples, she sobbed her heart out. Her heartbeat was in chaos. She was shivering. Her masseuse asked her to breathe deeply. She soon realized that the temple area where she received the massage was the same area which hurt when she had ophthalmic migraines in elementary school. During these terrible headaches, she would become blind for a few minutes. She was always concerned that her vision might not come back. Suddenly, she remembered that, in her childhood, she had once been punished by her elementary school teacher and was sent to a dark closet where she was scared to death. The fear she was experiencing in her head was hidden in her body.

Through her body, she opened a passage to access her Emotional Brain.

SINCE THE EMOTIONAL BRAIN (OLD BRAIN) IS MORE CONNECTED TO THE BODY THAN THE COGNITIVE BRAIN (NEW BRAIN), IT IS EASIER TO HAVE ACCESS TO THE EMOTIONS THROUGH THE HEART AND BODY.

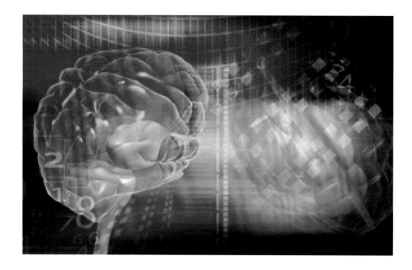

To create harmony between the two brains in the head and the brain in the heart, try the following Heart Harmony Exercise. This exercise brings coherence to the heart, thus opening the flow of information and energy in your Heart Brain system. Even though it brings about an inner calm, *it is not a relaxation technique.* It is meant to facilitate action. It can be practiced every day and almost anywhere. There are no special props necessary, and it only takes about ten minutes.

BASIC HEART HARMONY EXERCISE

Sit comfortably with a straight back to give freedom to the flow of air going from your nose to your lungs and your belly. You may feel more at ease lying down on a sofa with pillows behind your back.

Step One: Keep your worries briefly waiting or set aside your personal concerns. To accomplish the above, turn your attention inward. Put your hand on the area of your heart, between the pectoral muscles.

Step Two: Take three, slow, deep breaths with your belly. They will immediately stimulate the parasympathetic system and begin applying a bit of physiological "brake."

Step Three:
- Inhale slowly, count "1, 2, 3"
- Exhale slowly, count "5, 4, 3, 2, 1"
- Pause, count "1, 2"
- Repeat step three 5 times

During the exhalation, imagine that your breath travels through the heart. Breathing brings oxygen to the heart and expels carbon dioxide and waste. The heart is "washing" or purifying itself, and the body. Your attention must stay focused on your breathing until you have finished exhaling and pause. The intention is to let your mind float while exhaling, to the point where it (the mind) lightens up, becoming mellow and buoyant in your chest.

Let your breathing pause for a few seconds before the next inhalation (in breath) begins, and repeat 5 times slowly.

Step Four: GRATITUDE

Now, breathe normally, and **ENCOURAGE YOUR HEART WITH FEELINGS OF GRATITUDE AND LOVE FOR YOURSELF AND OTHERS.**

Then, imagine the face of a beloved child, a peaceful scene in nature, the memory of a physical feat, the perfect swing of a golf club, the smell of a flower, the wind caressing your skin (or whatever image brings you the greatest joy). You may notice a gentle smile on your lips, or a sensation of lightness, warmth, and expansion in your chest. This is the sign that harmony has been established.

Finally, allow the Old Brain and the New Brain to receive the peace of your heart. Imagine the peace ascending up your spinal column to your brain.

If you get distracted by thoughts such as, "What do I cook for my mother-in-law tonight?" Don't fight them. Let them surf away as you exhale, saying, "Later." The compelling thoughts will fade and disappear during the pause after the expiration (breathing out). Another one may come, "Do I put the blue tablecloth or the green one?" And another one, "Should I ask the children to dress properly?" With practice, these compelling thoughts will not come very often. Let them go like clouds pass by on a breezy day.

As we explained earlier, this exercise allows you to

- Control anxiety
- Lower blood pressure
- Increase the hormone DHEA (dehydroeprandrosterone)
- Stimulate the immune system
- Turn back the physiological clock
- Clear your mind
- Find your profound desires
- Have existing knowledge more readily accessible when in a stressful situation.
- Deal with insomnia
- Reduce stress
- Rid yourself of panic attacks
- Alleviate depression

The Heart Harmony Basic Exercise is brief and quickly adaptable to real life situations. The simplicity and the speed of this exercise works in virtually any context like the workplace, a store, public transportation, home, etc. You can use the exercise in the heat of battle. It operates in the emotional domain and directly transforms negative emotions, which are the source of most stress.

The Heart Harmony Basic Exercise is free of religious and cultural bias. It can be used by everybody. It addresses the cumulative, emotional, energy depletion that contributes to many illnesses.

THE HEART HARMONY BASIC EXERCISE INFUSES THE SYSTEM WITH PLEASURABLE, POSITIVE EMOTIONAL EXPERIENCES. YOU WILL IMMEDIATELY FEEL BETTER. There is no competition with any other therapy. As you have seen, this exercise acts via the heart (the heart rate harmony).

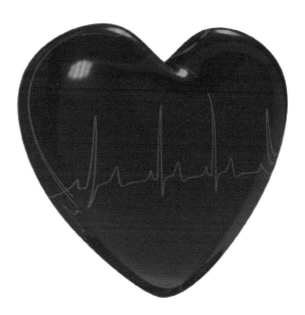

EMERGENCY HEART HARMONY EXERCISE

Sometimes your day is long and difficult and you take everything "to heart." You may become angry, frustrated, defensive or just tired. You may begin to say things that are negative, to react to situations in a disagreeable manner or to start eating food you don't need. The following Emergency Heart Harmony Exercise can be practiced in everyday situations, anywhere.

Step One: Focus your attention in the area around your heart. Place your hand over your heart if possible.

Step Two: Take two deep breaths. Imagine you are breathing through your heart, slowly and gently.

Step Three: Recall a positive feeling, a pleasant memory, while concentrating your attention on your heart.

This Emergency Heart Harmony Exercise interrupts the stress response and quickly brings your system into balance. It can be used:

- ♥ During emotionally challenging events or situations.

- ♥ To rebalance or to get an energy boost during the day.

- ♥ To maintain harmony during the day, practice in the morning, in the middle of the day, and at bedtime.

Both of the above exercises, the Heart Harmony Exercise and the Emergency Heart Harmony Exercise, have very practical applications:

- ♥ Stress reduction
- ♥ Emotional self-regulation
- ♥ Stress mediated illnesses
- ♥ Diminish chronic illnesses
- ♥ Decrease psycho-physiological conditions

These simple exercises can be used for many applications:

1. PHYSICAL PAIN: Negative emotional states, such as anxiety and fear, go hand in hand with physical pain. Therefore, these exercises may be used in reframing emotional states and in helping to reduce acute pain and chronic pain.

2. SOCIAL ISOLATION: Chronic pain may bring on isolation. Pain takes precedence over everything and leaves no energy for social interactions.

    Example: Edward had ophthalmic migraines with concurrent nausea. He used the Heart Harmony Exercise as an abortive deterrent when the migraine would begin and then later as a preventative measure.

3. PHYSIOLOGICAL DISORDERS

    ♥ Insomnia

    ♥ Irritable bowel syndrome

    ♥ Abdominal pain

    ♥ Headache

    ♥ Neurological Tics

    ♥ Hypertension

4. EMOTIONAL/BEHAVIORAL DISORDERS

    ♥ Anxiety

    ♥ Depression

    ♥ Anger

    ♥ Emotional instability

    ♥ ADHD (attention deficit hyperactivity disorder)

5. CHRONIC ILLNESSES

    ♥ Ulcerative Colitis

    ♥ Asthma

    ♥ Cystic Fibrosis

    ♥ Cancer

    ♥ Atopic Dermatitis

6. **SCHOLASTIC/ACADEMIC SUCCESS**
   Elementary, middle and high schools, as well as universities use the exercise to enhance learning.

7. **INTERACTION WITH OTHERS**
   At work with colleagues, with relatives, with friends, with your children ...

Example: During a work conference you are verbally criticized by your boss in front of your colleagues. Your body reacts by secreting great amounts of adrenaline in your blood, which cuts off access to your New Brain. You will probably either freeze or have a very negative reaction that could include words or actions that you may regret later. However, if instead you practice the Emergency Heart Harmony Exercise, you will have different ways to react. You will have access to your New Brain, and find a way to behave that will not compromise your boss's ego and will save you from even more ridicule.

For more help in practicing these exercises, please refer to "The Practice of Heart Harmony" on the CD attached to the inside back cover of this book.

CHAPTER 13

Heart Harmony Conclusions

To promote harmony in your life or to recover it when it has been lost, practice the Heart Harmony Exercises. You may use them daily or as needed.

First listen to your heart, and by applying the Heart Harmony Basic Exercise, you can control your cardiac rhythm. The good news is that *all of you*, *at any age*, are capable of governing your emotions and your relationships with others.

Remember all the benefits of the Heart Harmony Exercises:

- ♥ Withstand stress
- ♥ Control anxiety
- ♥ Maximize survival energy
- ♥ Slow down aging
- ♥ Achieve peace of mind

As you recall, stress is not caused by outside events. It is actually caused by emotional reactions to events. By maintaining a genuine connection with your heart you will gain access to peace, a higher quality of life, and happiness.

Your body and especially your heart are precious allies whose messages never mislead you. They give you notice of what is in your best interest or what does not serve you. And through the heart, you can cope with difficult feelings and give yourself a new perspective.

THROUGH THE REPETITION OF PRACTICING THE HEART HARMONY EXERCISES YOU WILL LEARN TO TRUST YOUR HEART, THEREBY ACHIEVING GREATER HAPPINESS.

These exercises give an added tool to medication. They are safe and are also non-pharmacological.

Practicing these focused exercises teaches you to breathe comfortably in stressful situations (aerobic physical activity and charged emotional states).

Despite ongoing stresses, you are able to better handle challenging situations by:

- ♥ Focusing on the positive aspect of the situation
- ♥ Controlling insomnia
- ♥ Managing frustration
- ♥ Protecting interpersonal relationships.

The Heart Harmony Exercises are tools for stress reduction, emotional self-regulation and self-control. They can be used to treat most stress medicated illnesses, many chronic illnesses and conditions that are psycho-physiological in nature. Our heart rate is sensitive to different events taking place in our mind and body.

These two exercises lead to inner calm. They are not conventional relaxation techniques. They facilitate action in a stressful or challenging situation, allowing you an opportunity to reach for your inner resources.

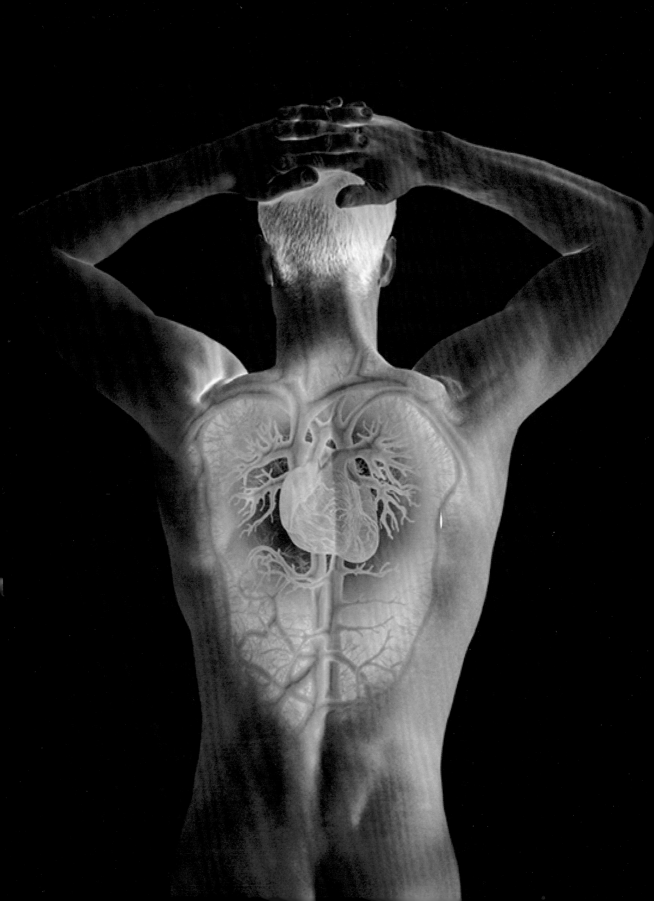

CHAPTER 14

Final Conclusions

Your perceptions and your emotions were once believed to be dictated entirely by the responses of the New Brain to stimuli coming from your outside environment. **TODAY, SCIENTISTS DESCRIBE PERCEPTIONS AND EMOTIONAL EXPERIENCES AS A COMBINATION OF THE STIMULI RECEIVED BY THE OLD BRAIN FROM THE ORGANS IN THE BODY.** The brain in the head—once thought of as a unitary organ—is in reality two separate brains, vastly different from one another in both structure and function—the Old Brain (the emotional brain) and the New Brain (the cognitive brain).

The heart plays a central role in the generation of emotional experiences and also in the establishment of psychological and physical harmony, sense-of-purpose and wellbeing. **IF YOU DON'T CONTROL YOUR EMOTIONS, YOUR BODY SUFFERS AND YOU ARE STRESSED.** Stress affects you physically, mentally and emotionally.

To achieve wellbeing the Heart Brain must be in harmony and there must be cooperation between the two brains in the head (the Old Brain and the New Brain). When there is no cooperation, chaos occurs. If the Old Brain is out of order, the heart suffers. If the heart does not function at its best, it negatively influences the Old Brain, which shuts down and does not transfer needed information to the New Brain.

The intellect cannot resolve a dysfunction between the heart and the brain in the head by attempting to control emotions. **IT IS BY LEARNING TO LISTEN TO AND ACT FROM YOUR HEART AND BRAIN IN SYNCHRONY THAT YOU HAVE A MORE PRODUCTIVE, HEALTHY AND SATISFYING LIFE.** To establish this synchrony you need to reach Heart Harmony. How? Practice the Heart Harmony Exercises. These exercises have a direct impact on the Emotional Brain (Old Brain).

The harmony you will achieve will be beneficial to your body and useful for mastering your emotions.

You may gain more conscious control over your mental and your emotional states. **BE PREPARED TO FIND A NEW ZEST FOR LIVING, MORE MEANING AND A REAL CONNECTION WITH YOUR INNATE JOY.**

*It's really so simple.
Just go to your heart and breathe!*

ACKNOWLEDGMENTS

We would like to take this opportunity to thank our wonderful families and gracious friends for their continued support of this project.

We would especially like to thank Dotti Albertine of Albertine Book Design, for her excellent skills in editing and her brilliance in book design.

END NOTES

FOREWORD
1. http://www.jasonlincolnjeffers.com/circuitry.html

2. Kresh, J. Yasha; Armour, J. Andrew; "The heart as a self-regulating system: integration of homeodynamic mechanisms", Technology Health Care, Volume 5

INTRODUCTION
1. Servan-Schreiber, David; *Guérir le Stress (The Instinct to Heal)*

CHAPTER ONE
1. *Webster's Encyclopedic Unabridged Dictionary*

CHAPTER TWO
1. Carlson, Neil R.; *Physiology of Behavior, 10th Edition*

2. Levitt, P; *A Monoclonal Antibody to Lymbic System Neurons,* "*Science* 223" (1984): 299-301.

3. Damasio, Antonio; *The Feeling of What Happens*

4. Goleman, Daniel; *Emotional Intelligence*

CHAPTER FOUR
1. LeDoux, Joseph; *The Emotional Brain*

2. Carlson, Neil R.; *Physiology of Behavior, 10th Edition*

3. Heart-Math Institute; "Tools & Techniques"

4. Servan-Schreiber, David; *Guérir le Stress (The Instinct to Heal)*

CHAPTER FIVE
1. Servan-Schreiber, David; *Guérir le Stress (The Instinct to Heal)*

CHAPTER EIGHT
1. Oschman, James L. (1997). "What is healing energy? Part 3: silent pulses." *Journal of Bodywork and Movement Therapies 1* (3): 179–189. doi:10.1016/S1360-8592(97)80038-1

2. Pawluk, William MD. M.Sc., www.drpawluk.com/health/electromagnetic-fields-and-the-heart.

3. Shinn, F. S. (1925). *The Game of Life. In The Writings of Florence Scovel Shinn* (1988, pp. 1 – 92). Marina del Rey, CA: DeVorss & Company.

4. Braden, Gregg, "Our ElectroMagnetic HEART Affects Reality: https://www.youtube.com/watch?v=X1SMqQH7FIU

5. Emoto, Dr. Masaru. The Message from Water: *The Message from Water is Telling Us to Take a Look at Ourselves 1*. Hado. 2000. ISBN 9784939098000.

6. Braden, Gregg, Ibid.

7. Pawluk, William MD. M.Sc., *Electromagnetic Fields in the Heart: Basic Science and Clinical Use*, December 2003

8. McCraty R, Atkinson M, Tomasino D, Bradley RT. *The Coherent Heart: Heart-Brain Interactions, Psychophysiological Coherence, and the Emergence of System-Wide Order*. Boulder Creek, CA: HeartMath Research Center, Institute of HeartMath, Publication No. 06-022; 2006.

9. Pribram, Karl (1971). *Languages of the brain; experimental paradoxes and principles in neuropsychology*. Englewood Cliffs, N.J.: Prentice-Hall. ISBN 0-13-522730-5.

10. McCraty R, Atkinson M, Tomasino D, Bradley RT, Ibid.

Chapter Nine
1. Servan-Schreiber, David; *Guérir le Stress (The Instinct to Heal)*

Chapter Ten
1. Servan-Schreiber, David; *Guérir le Stress (The Instinct to Heal)*

2. Servan-Schreiber, David; *Guérir le Stress (The Instinct to Heal)*

3. *Webster's International Dictionary of the English Language*

4. Seligman, Martin E. P.; *Learned Optimism: How to Change Your Mind and Your Life*

Chapter Eleven
1. Dreher, Henry; *Mind-Body Unity: A New Vision for Mind-Body Medicine*

2. Seligman, Martin E. P.; *Learned Optimism: How to Change Your Mind and Your Life*

3. Stanford University School of Medicine; Stanford, California

4. Servan-Schreiber, David; *Guérir le Stress (The Instinct to Heal)*

Chapter Twelve
1. Dreher, Henry; *Mind-Body Unity: A New Vision for Mind-Body Medicine*

GLOSSARY

Acetylcholine: Hormone that promotes states of relaxation, slows down the heart beat and calms the Old Brain.

Adrenaline and Noradrenaline: Hormones that speed up the heart and activate the Old Brain.

ANF (atrial natriuretic factor): Hormone that regulates blood pressure.

Autocompletion: When a life form strives towards order, coherence and purity.

Autonomic peripheral nervous system: A mechanism which connects the heart with the brain (in the head). It has two branches.

Body Coherence: When a system is coherent, virtually no energy is wasted (because of the internal synchronization among the parts).

> Example: In a firm, when everyone works in coherence—positive cooperation—creativity and productivity emerge at all levels.

Cardiac Coherence: When the heart's rhythmic beating is highly ordered. The heart beating alternates regularly between speeding up and slowing down.

Cortical brain: Also known as Cognitive Brain and New Brain. It controls cognition, language and reasoning.

Cortisol: hormone responsible for stress.

DHEA (dehydroepiandrosterone): This hormone is called the "youth" hormone.

Emotion: A strong feeling such as love, joy, anger …

Emotional Intelligence: Capacity to identify your emotional state and the emotional state of others. Mental capacities by which you can predict an individual's success.

Happiness: Wellbeing, satisfaction to be where you are at the present moment.

Harmony: A state when the Heart and the Emotional Brain are connected and there is an agreement between both of them.

Heart: Muscular organ which contracts and relaxes to keep the blood circulating. It has its "little brain" which communicates with the brain in the head.

Heart Intelligence: The heart is an intelligent system with the power to bring both the emotional and mental systems into balance and coherence.

Heart Rate Variability (HRV): It is the number of heart beats per minute. The normal variability swings (oscillates) from 55 to 70 beats per minute with an average of 62 beats per minute. The HRV is an important tool because it is a key indicator of cardiac health and wellbeing.

Homeostasis: A state of harmony among all physical functions. The dynamic balance or equilibrium between all functions in the body that keeps us alive.

IgA (immunoglobulin A): responsible for resistance to infection.

Lifeforce: Desire to live. Aristotle speaks about the need for self-fulfillment that starts with the seed and comes to full fruition in the tree.

Limbic brain: also known as the Emotional Brain and the Old Brain. It controls emotions and the body's physiology.

Oxytocin (love peptide): Hormone acting during breastfeeding, courtship and orgasm.

Parasympathetic branch: Acts as a brake on the heart and the Emotional Brain. It releases acetylcholine.

Perception: The way we view a situation or event.

PTSD: post-traumatic stress disorder.

Self healing: Capacity to recover equilibrium and wellbeing.

Stress: State of negative emotional arousal, usually associated with feelings of discomfort or anxiety that we attribute to outside unpleasant events and which are in reality our emotional reactions to events.

Symbiosis: Close bond between the two brains (Old Brain and New Brain) which need each other to function. The association of the two brains—the Old Brain and the New Brain—is mutually beneficial to their functioning.

Sympathetic branch: Speeds up the heart and activates the Emotional Brain. It releases adrenaline and noradrenaline.

REFERENCES

David Servan-Schreiber, M.D., Ph.D. Rodal Inc, USA: *The Instinct to Heal.* This book was originally published in French, as "*Guérir le stress, l'anxiété et la dépression sans médicaments ni psychanalyse*" by Editions Robert Laffont, S.A., Paris, 2003

Institute of HeartMath
14700 West Park Avenue
Boulder Creek, CA 95006
www.heartmath.com